Contents

About *Science Centers, Grades PreK–K* 2

Making a File Folder Center 3

Living or Not Living? (Identifying Living and Nonliving Things) 4

Where Is It? (Relative Position of Objects) 14

3 of a Kind (Compare and Sort by Physical Attributes) 26

Plant or Animal? (Classifying Plants and Animals) 38

Animal Families (Compare and Sort by Physical Attributes) 52

A Sunflower Grows (Understanding Plant Life Cycles) 62

A Chicken Grows (Understanding Animal Life Cycles) 72

How Many Legs? (Identifying Major Structures of Animals) 82

Parts of a Plant (Identifying Major Structures of a Plant) 96

Let's Eat! (Understanding Where Food Comes From) 106

My Body (Identifying Human Body Parts) 116

What Will I Wear Today? (Understanding How Weather Affects Our Lives) 126

Animal Homes (Identifying Animal Habitats) 144

Solid or Liquid? (Recognizing Liquid and Solid Forms) 154

Animal Puzzles (Identifying Major Structures of Animals) 168

Sun, Earth, Moon (Identifying Sun, Earth, and Moon) 180

Answer Key 190

About Science Centers
Grades PreK–K

What's Great About This Book

Centers are a wonderful, fun way for students to practice important skills. The 16 centers in this book are self-contained and portable. Students may work at a desk, at a table, or even on the floor. Once you've made the centers, they're ready to use any time.

What's in This Book

The teacher directions page includes how to make the center and a description of the student task

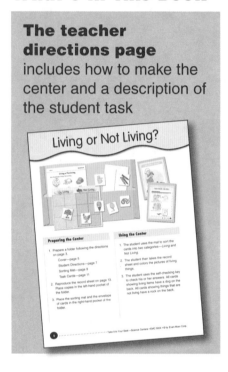

Full-color materials needed for the center

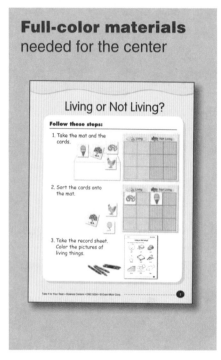

Reproducible activity sheets

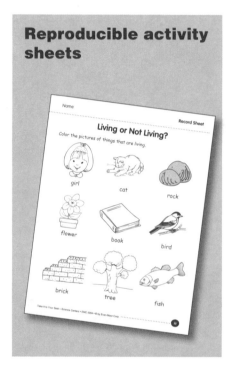

How to Use the Centers

The centers are intended for skill practice, not to introduce skills. It is important to model the use of each center before students perform the task independently.

Questions to Consider:

- Will students select a center, or will you assign the centers?
- Will there be a specific block of time for centers, or will the centers be used throughout the day?
- Where will you place the centers for easy access by students?
- What procedure will students use when they need help with the center tasks?
- Where will students store completed work?
- How will you track the tasks and centers completed by each student?

Making a File Folder Center

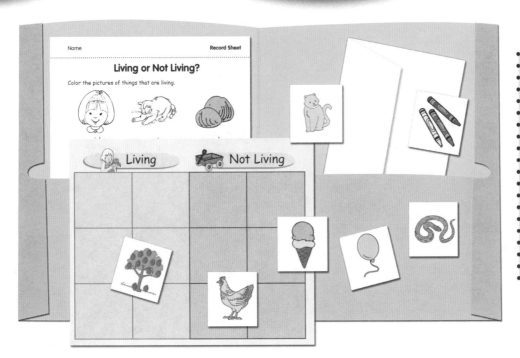

Folder centers are easily stored in a box or file crate. Students take a folder to their desks to complete the task.

Materials

- folder with pockets
- envelopes
- marking pens and pencils
- scissors
- stapler
- two-sided tape

Steps to Follow

1. Laminate the cover. Tape it to the front of the folder.
2. Laminate the student directions page. Tape it to the back of the folder.
3. Reproduce the record sheet. Place record sheets, writing paper, and any other supplies in the left-hand pocket of the folder.
4. Laminate the task cards. Place each set of task cards in an envelope. Place the labeled envelopes in the right-hand pocket of the folder.
5. Laminate the sorting mat(s). Place the mat(s) in the right-hand pocket of the folder.

Take It to Your Seat—Science Centers • EMC 5004 • © Evan-Moor Corp.

Living or Not Living?

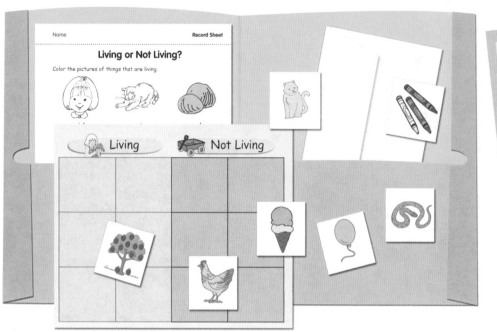

Preparing the Center

1. Prepare a folder following the directions on page 3.
    - Cover—page 5
    - Student Directions—page 7
    - Sorting Mat—page 9
    - Task Cards—page 11

2. Reproduce the record sheet on page 13. Place copies in the left-hand pocket of the folder.

3. Place the sorting mat and the envelope of cards in the right-hand pocket of the folder.

Using the Center

1. The student uses the mat to sort the cards into two categories—*Living* and *Not Living*.

2. The student uses the self-checking cards to check his or her answers. All cards showing living items have a girl on the back. All cards showing things that are not living have a wagon on the back.

3. The student then takes the record sheet and colors the pictures of living things.

Living or Not Living?

The girl is living.

The wagon is not living.

Living or Not Living?

Follow these steps:

1. Take the cards and the mat.

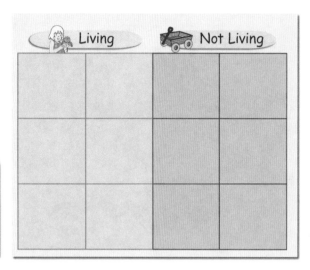

2. Sort the cards onto the mat.

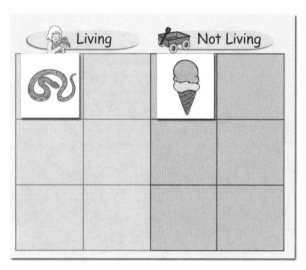

3. Take the record sheet. Color the pictures of living things.

Not Living

Living

© Evan-Moor Corp. © Evan-Moor Corp. © Evan-Moor Corp.

© Evan-Moor Corp. © Evan-Moor Corp. © Evan-Moor Corp.

© Evan-Moor Corp. © Evan-Moor Corp. © Evan-Moor Corp.

© Evan-Moor Corp. © Evan-Moor Corp. © Evan-Moor Corp.

Living or Not Living?

Color the pictures of things that are living.

Where Is It?

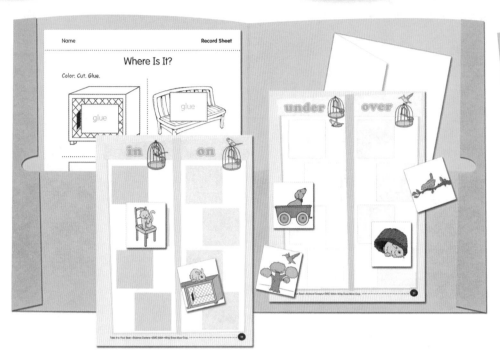

Preparing the Center

1. Prepare a folder following the directions on page 3.

    Cover—page 15

    Student Directions—page 17

    Sorting Mats—pages 19 and 21

    Task Cards—page 23

2. Reproduce the record sheet on page 25. Place copies in the left-hand pocket of the folder.

3. Place the sorting mats and the envelope of cards in the right-hand pocket of the folder.

Using the Center

1. The student sorts the cards into four sets on the mats—*in, on, under,* and *over*.

2. The student uses the self-checking cards to check his or her answers. All four cards belonging together have a square on the back in the same color as the correct mat.

3. Then the student cuts out the animals on the record sheet and glues them in the correct place.

Where Is It?

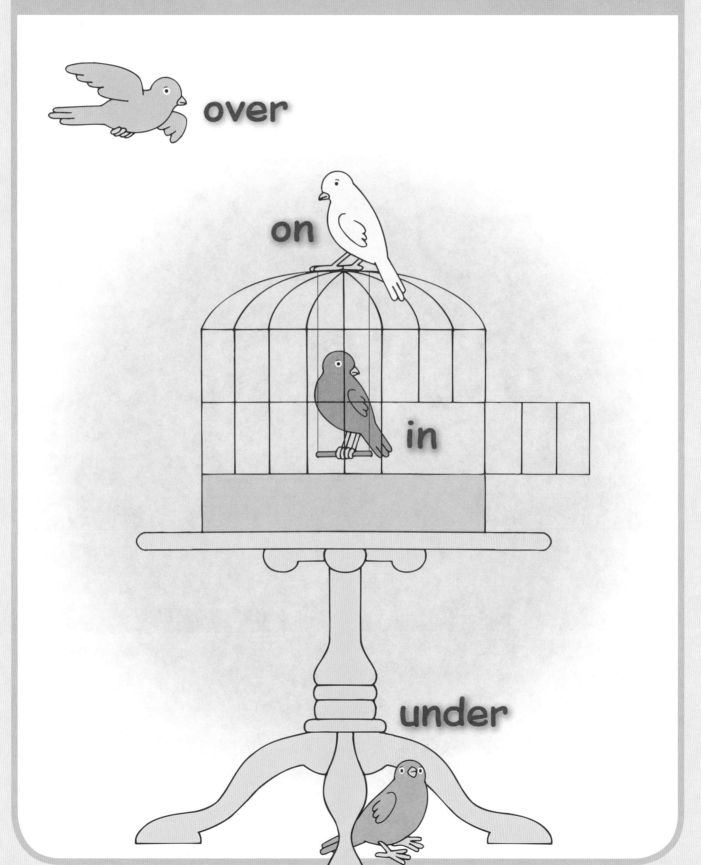

Where Is It?

Follow these steps:

1. Take the cards and the mats.

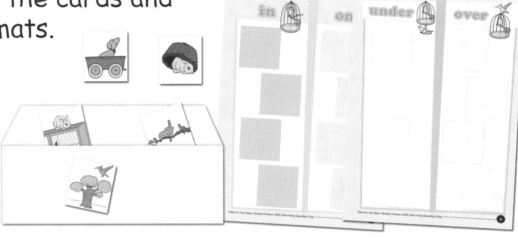

2. Sort the cards onto the mats.

  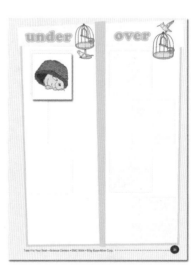

3. Take the record sheet. Color. Cut. Glue.

Take It to Your Seat—Science Centers • EMC 5004 • © Evan-Moor Corp.

under  over

Name _____   Record Sheet

Where Is It?

Color. Cut. Glue.

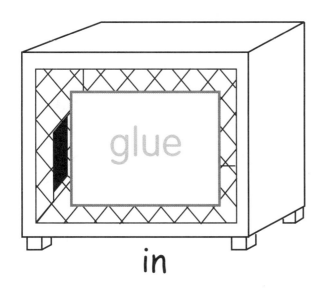
in

on

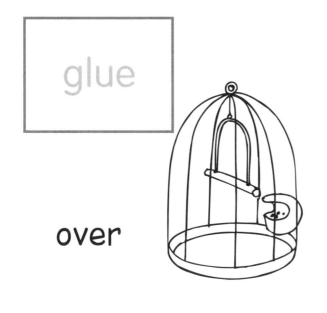
over

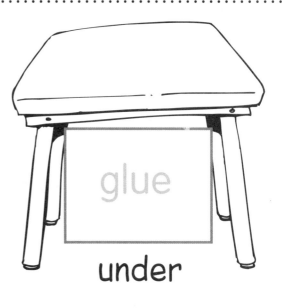
under

3 of a Kind

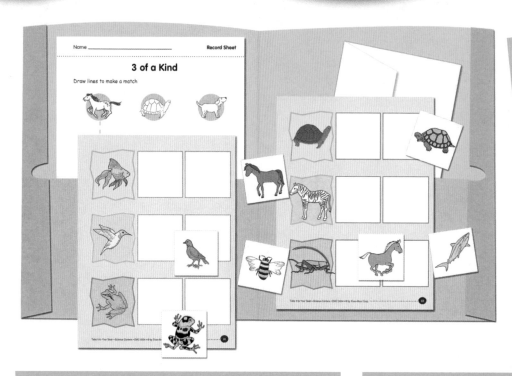
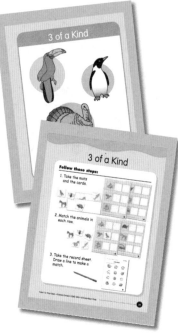

Preparing the Center

1. Prepare a folder following the directions on page 3.

    Cover—page 27

    Student Directions—page 29

    Sorting Mats—pages 31 and 33

    Task Cards—page 35

2. Reproduce the record sheet on page 37. Place copies in the left-hand pocket of the folder.

3. Place the sorting mats and the envelope of cards in the right-hand pocket of the folder.

Using the Center

1. The student sorts the cards, matching the same type of animal, bird, insect, or fish on the sorting mats.

2. The student uses the self-checking cards to check his or her answers. Students turn over the two cards they placed in each row on each sorting mat. Correctly matched cards will make a row of three identical pictures, for example, three fish, three turtles, etc.

3. Then the student draws lines to make a match on the record sheet.

3 of a Kind

We are all birds.

3 of a Kind

Follow these steps:

1. Take the cards and the mats.

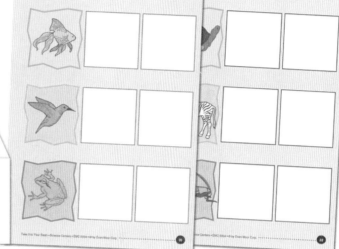

2. Match the animals in each row.

3. Take the record sheet. Draw a line to make a match.

 © Evan-Moor Corp.
 © Evan-Moor Corp.
 © Evan-Moor Corp.

 © Evan-Moor Corp.
 © Evan-Moor Corp.
 © Evan-Moor Corp.

 © Evan-Moor Corp.
 © Evan-Moor Corp.
 © Evan-Moor Corp.

 © Evan-Moor Corp.
 © Evan-Moor Corp.
 © Evan-Moor Corp.

Take It to Your Seat—Science Centers • EMC 5004 • © Evan-Moor Corp.

Name _____ Record Sheet

3 of a Kind

Draw lines to match 3 of a kind.

Plant or Animal?

Preparing the Center

1. Prepare a folder following the directions on page 3.

    Cover—page 39

    Student Directions—page 41

    Sorting Mats—pages 43 and 45

    Task Cards—pages 47 and 49

2. Reproduce the record sheet on page 51. Place copies in the left-hand pocket of the folder.

3. Place the sorting mats and the envelope of cards in the right-hand pocket of the folder.

Using the Center

1. The student sorts the cards into two sets on the mats—*Plants* and *Animals*.

2. The student uses the self-checking cards to check his or her answers. All cards showing animals have a koala on the back. All cards showing plants have a leaf on the back.

3. Then the student cuts and glues the plants and animals into the correct boxes on the record sheet.

Plant or Animal?

Plant or Animal?

Follow these steps:

1. Take the cards and the mats.

2. Sort the cards onto the mats.

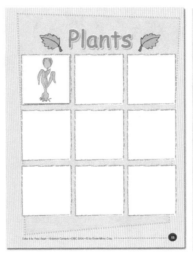

3. Take the record sheet. Color. Cut. Glue.

Take It to Your Seat—Science Centers • EMC 5004 • © Evan-Moor Corp.

© Evan-Moor Corp.   © Evan-Moor Corp.   © Evan-Moor Corp.

© Evan-Moor Corp.   © Evan-Moor Corp.   © Evan-Moor Corp.

© Evan-Moor Corp.   © Evan-Moor Corp.   © Evan-Moor Corp.

 © Evan-Moor Corp.  © Evan-Moor Corp.  © Evan-Moor Corp.

 © Evan-Moor Corp.   © Evan-Moor Corp.  © Evan-Moor Corp.

 © Evan-Moor Corp.   © Evan-Moor Corp.  © Evan-Moor Corp.

Name _____   Record Sheet

Plant or Animal?

Color. Cut. Glue.

Plants

Animals

Animal Families

Preparing the Center

1. Prepare a folder following the directions on page 3.

    Cover—page 53

    Student Directions—page 55

    Puzzle Pieces—pages 57 and 59

2. Reproduce the record sheet on page 61. Place copies in the left-hand pocket of the folder.

3. Place the envelope of puzzle pieces in the right-hand pocket of the folder.

Using the Center

1. The student puts the puzzle pieces together, matching the animal parents to their animal babies.

2. The student uses the self-checking puzzle pieces to check his or her answers. Matching pairs have the same color shape on the back.

3. Then the student cuts and glues to match the animal parents and babies on the record sheet.

Animal Families

Animal Families

Follow these steps:

1. Take the puzzle pieces.

2. Match two pieces.

3. Take the record sheet. Color. Cut. Glue.

 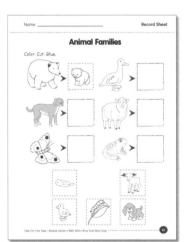

Take It to Your Seat—Science Centers • EMC 5004 • © Evan-Moor Corp.

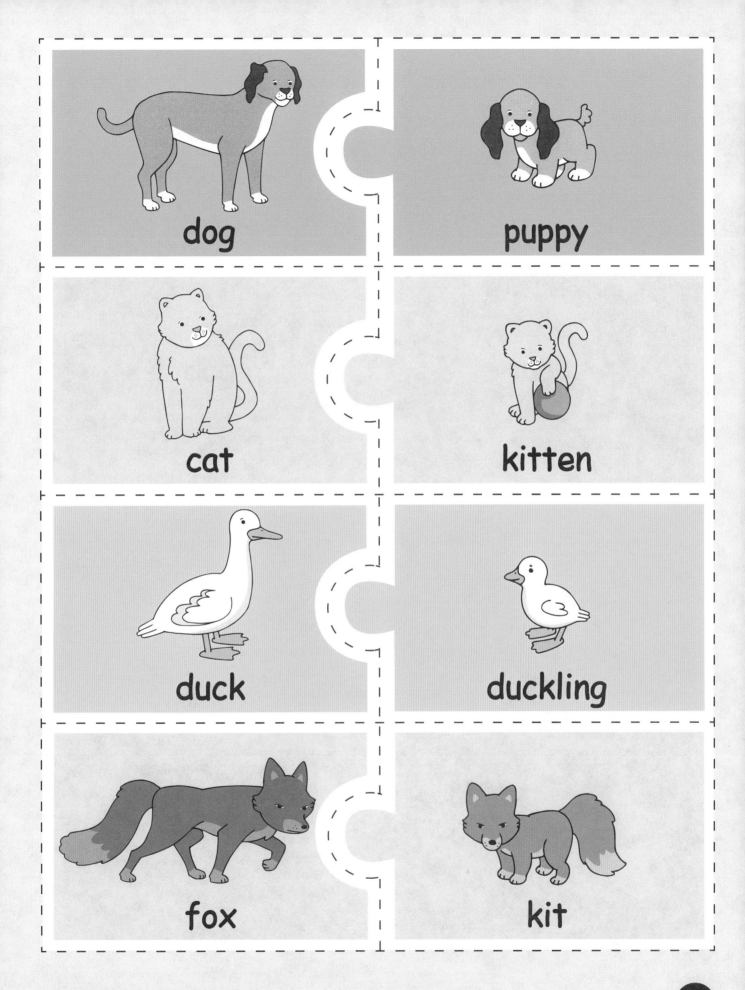

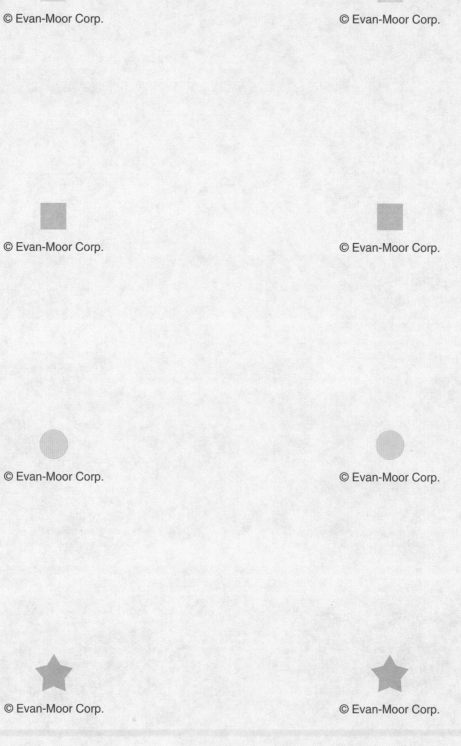

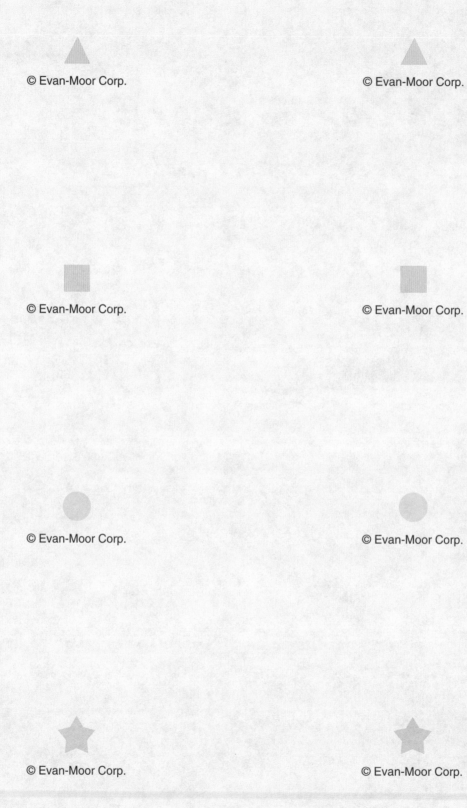

Name _____ Record Sheet

Animal Families

Color. Cut. Glue.

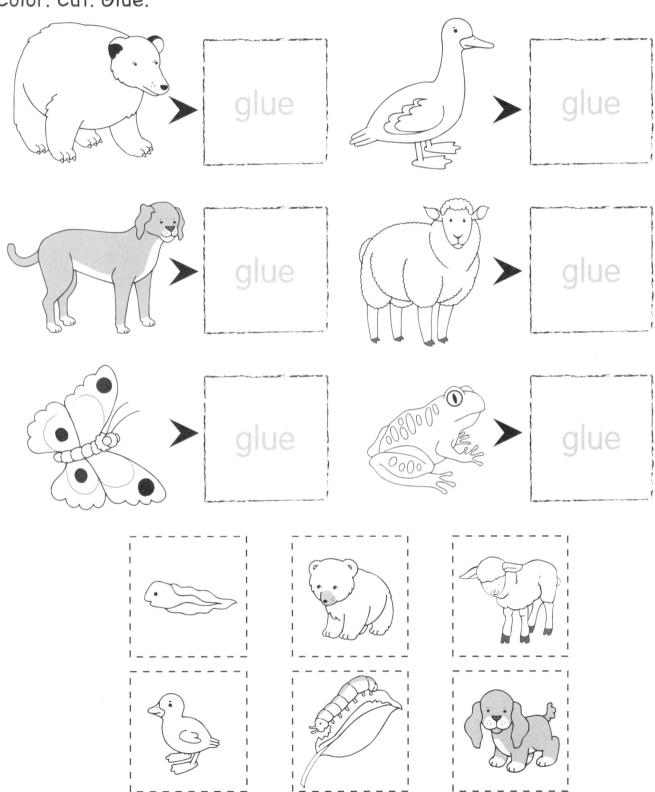

A Sunflower Grows

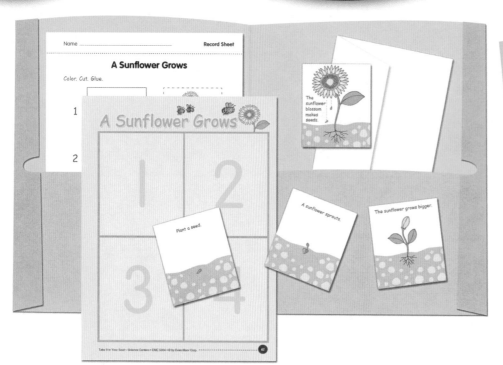

Preparing the Center

1. Prepare a folder following the directions on page 3.

    Cover—page 63

    Student Directions—page 65

    Sorting Mat—page 67

    Task Cards—page 69

2. Reproduce the record sheet on page 71. Place copies in the left-hand pocket of the folder.

3. Place the sorting mat and the envelope of cards in the right-hand pocket of the folder.

Using the Center

1. The student takes the sorting mat and cards, and then sequences the life cycle of the sunflower.

2. The student uses the self-checking cards to check his or her answers. The student turns the cards over. When in the correct order, the cards will form a sunflower in bloom.

3. Using the record sheet, the student then cuts and glues to sequence the life cycle of the sunflower.

A Sunflower Grows

A Sunflower Grows

Follow these steps:

1. Take the cards and the mat.

2. Put the cards in order.

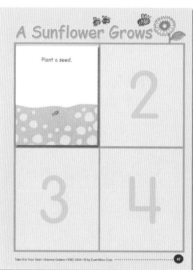

3. Take the record sheet. Color. Cut. Glue.

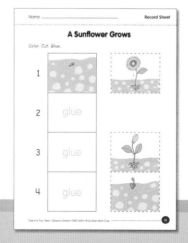

A Sunflower Grows

1	2
3	4

A sunflower sprouts.

Plant a seed.

The sunflower blossom makes seeds.

The sunflower grows bigger.

Name _____   Record Sheet

A Sunflower Grows

Color. Cut. Glue.

1 glue

2 glue

3 glue

4 glue

Take It to Your Seat—Science Centers • EMC 5004 • © Evan-Moor Corp.   71

A Chicken Grows

Preparing the Center

1. Prepare a folder following the directions on page 3.

    Cover—page 73

    Student Directions—page 75

    Sorting Mat—page 77

    Task Cards—page 79

2. Reproduce the record sheet on page 81. Place copies in the left-hand pocket of the folder.

3. Place the sorting mat and the envelope of cards in the right-hand pocket of the folder.

Using the Center

1. The student takes the sorting mat and cards, and then sequences the life cycle of the chicken.

2. The student uses the self-checking cards to check his or her answers. The student turns the cards over. When in the correct order, the cards form a hen.

3. Using the record sheet, the student then cuts and glues to sequence the life cycle of the chicken.

A Chicken Grows

A Chicken Grows

Follow these steps:

1. Take the cards and the mat.

2. Put the cards in order.

3. Take the record sheet. Color. Cut. Glue.

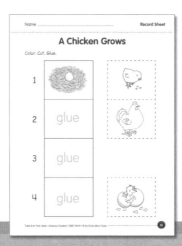

A Chicken Grows

1	2
3	4

Take It to Your Seat—Science Centers • EMC 5004 • © Evan-Moor Corp.

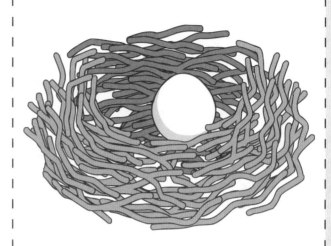

The chick hatched out of the shell.

An egg is in the nest.

The hen is grown.
Now she can lay eggs.

The chicken is growing.

Name _____ Record Sheet

A Chicken Grows

Color. Cut. Glue.

1	glue
2	glue
3	glue
4	glue

How Many Legs?

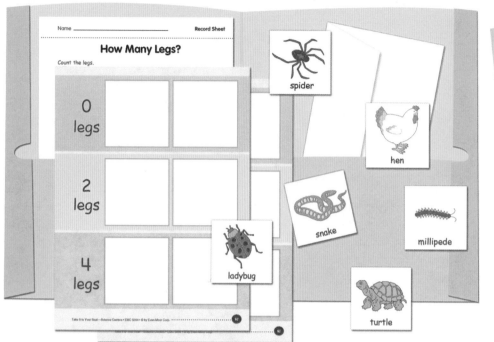

Preparing the Center

1. Prepare a folder following the directions on page 3.

    Cover—page 83

    Student Directions—page 85

    Sorting Mats—pages 87 and 89

    Task Cards—pages 91 and 93

2. Reproduce the record sheet on page 95. Place copies in the left-hand pocket of the folder.

3. Place the sorting mats and the envelope of cards in the right-hand pocket of the folder.

Using the Center

1. The student takes the sorting mats and cards, and then sorts the cards by number of legs.

2. The student uses the self-checking cards to check his or her answers. The number of legs is recorded on the back of each card.

3. Then the student completes the record sheet by counting and recording the number of legs of various animals.

How Many Legs?

How Many Legs?

Follow these steps:

1. Take the cards and the mats.

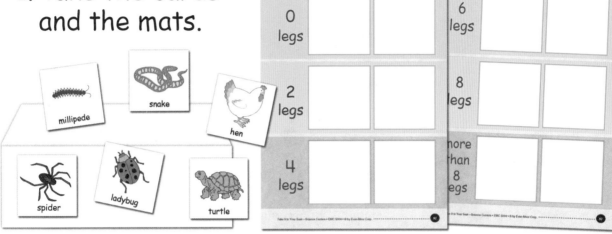

2. Sort the cards onto the mats.

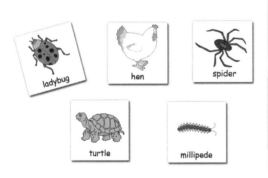
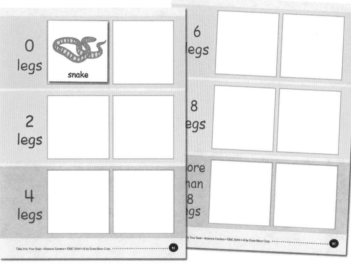

3. Take the record sheet.
   Count the legs.
   Write the number.
   Color.

0 legs

2 legs

4 legs

6 legs

8 legs

more than 8 legs

© Evan-Moor Corp.

© Evan-Moor Corp.

© Evan-Moor Corp.

© Evan-Moor Corp.

© Evan-Moor Corp.

© Evan-Moor Corp.

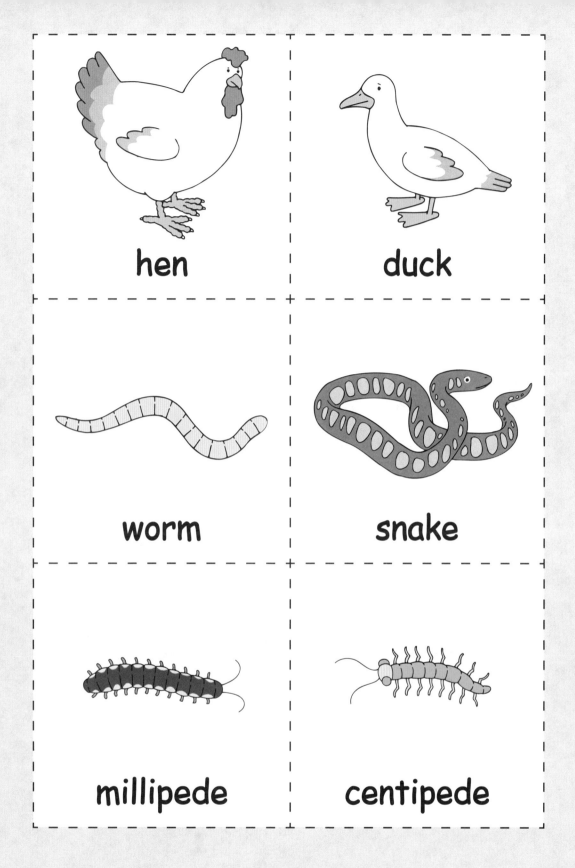

© Evan-Moor Corp.

© Evan-Moor Corp.

© Evan-Moor Corp.

© Evan-Moor Corp.

© Evan-Moor Corp.

© Evan-Moor Corp.

Name _____   Record Sheet

How Many Legs?

Count the legs.
Write the number.
Color.

I have _____ legs.

I have _____ legs.

I have _____ legs.

I have _____ legs.

I have _____ legs.

I have _____ legs.

Parts of a Plant

Preparing the Center

1. Prepare a folder following the directions on page 3.

    Cover—page 97

    Student Directions—page 99

    Puzzle Pieces—pages 101 and 103

2. Reproduce the record sheet on page 105. Place copies in the left-hand pocket of the folder.

3. Place the envelope of puzzle pieces in the right-hand pocket of the folder.

Using the Center

1. The student takes the puzzle pieces and puts them together to make a tree and a flower.

2. Correctly completed puzzles will show a complete tree or a flower.

3. Then the student completes the record sheet by cutting and gluing the missing parts in place.

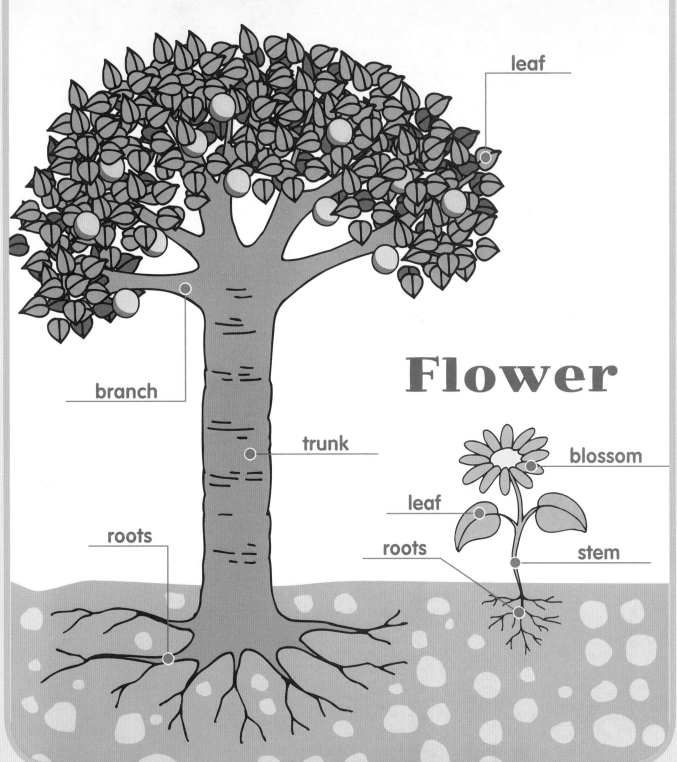

Parts of a Plant

Follow these steps:

1. Take the puzzle pieces.

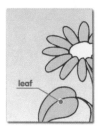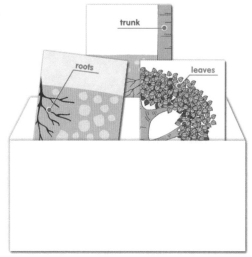

2. Make a tree and a flower.

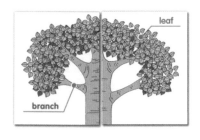

3. Take the record sheet. Color. Cut. Glue.

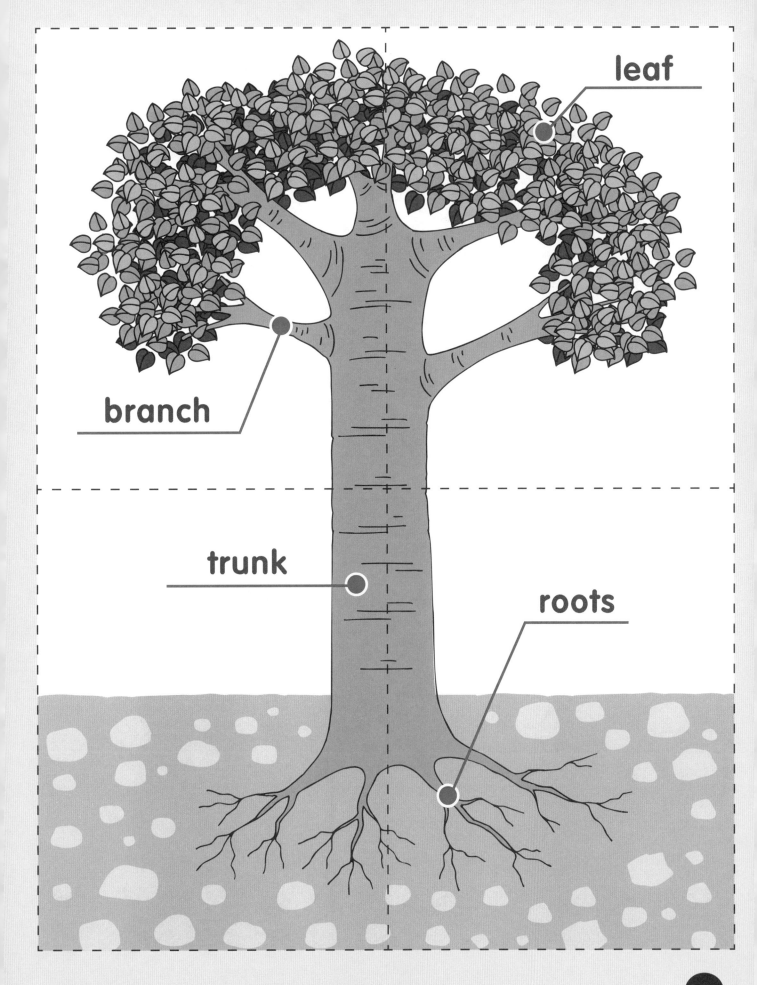

© Evan-Moor Corp.

© Evan-Moor Corp.

© Evan-Moor Corp.

© Evan-Moor Corp.

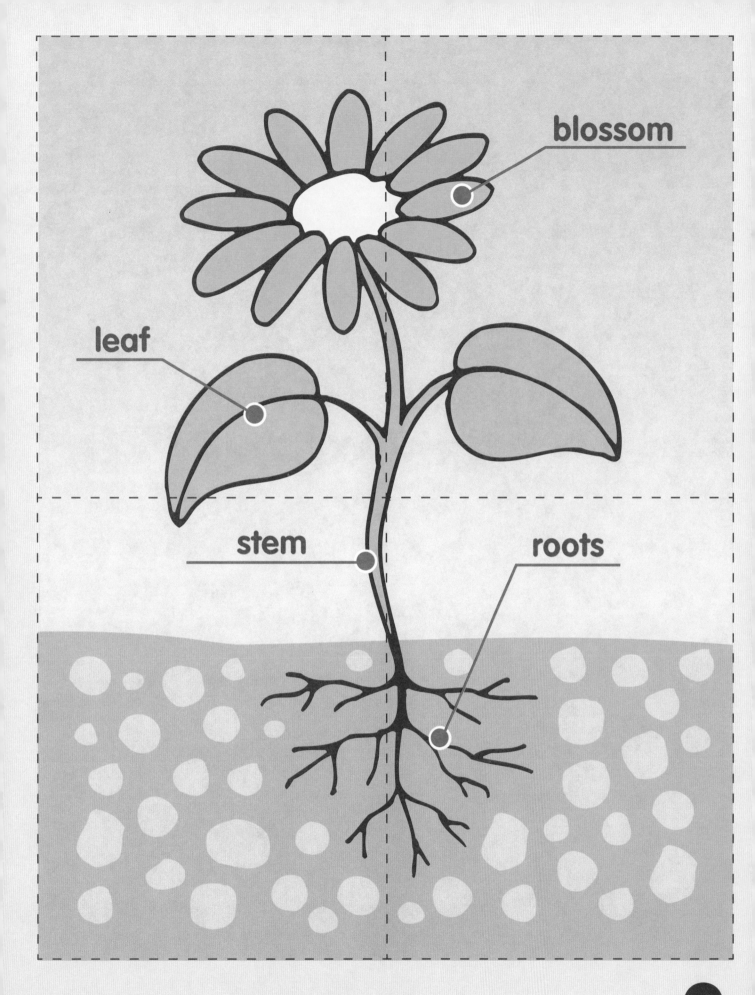

© Evan-Moor Corp.

© Evan-Moor Corp.

© Evan-Moor Corp.

© Evan-Moor Corp.

Name _____   Record Sheet

Parts of a Plant

Color. Cut. Glue.

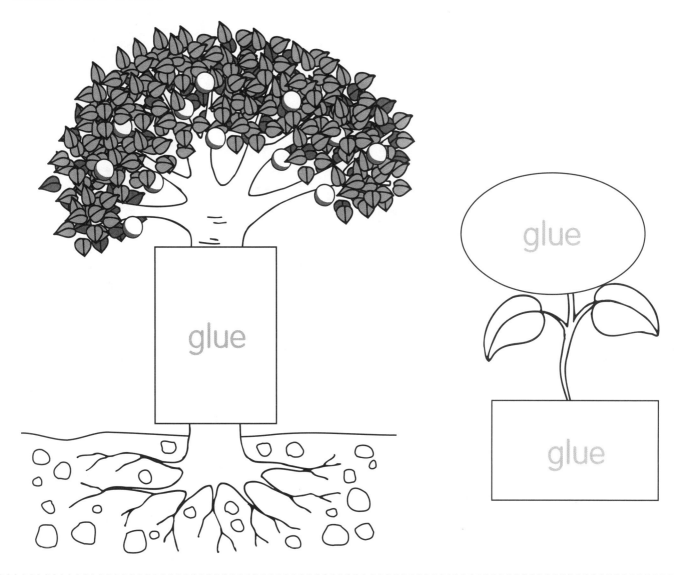

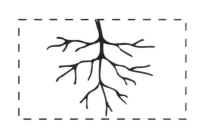
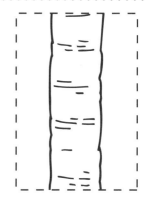

Take It to Your Seat—Science Centers • EMC 5004 • © Evan-Moor Corp.   105

Let's Eat!

Preparing the Center

1. Prepare a folder following the directions on page 3.

    Cover—page 107

    Student Directions—page 109

    Puzzle Pieces—pages 111 and 113

2. Reproduce the record sheet on page 115. Place copies in the left-hand pocket of the folder.

3. Place the envelope of puzzle pieces in the right-hand pocket of the folder.

Using the Center

1. The student puts the puzzle pieces together, matching the plant or animal with the food it provides.

2. The student uses the self-checking puzzle pieces to check his or her answers. Matching pieces have the same color shape on the back.

3. Then the student completes the record sheet by drawing a line to match each plant or animal with the food it provides.

Let's Eat!

Let's Eat!

Follow these steps:

1. Take the puzzle pieces.

2. Match two pieces.

3. Take the record sheet. Draw a line to make a match.

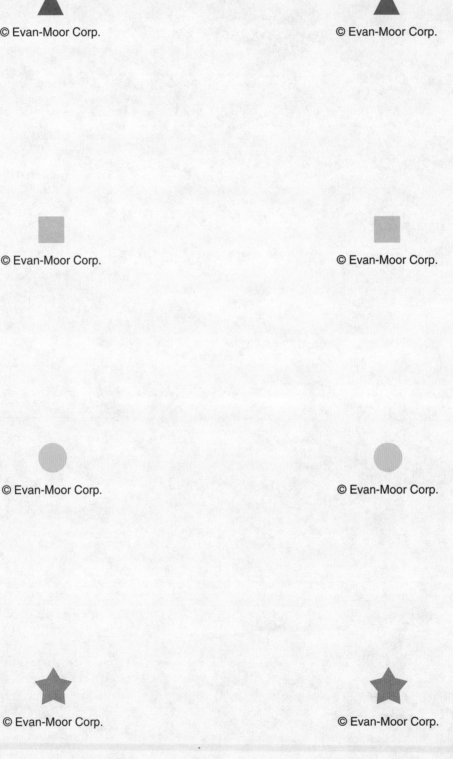

Name _____   **Record Sheet**

Let's Eat!

Draw a line to make a match.

My Body

Preparing the Center

1. Prepare a folder following the directions on page 3.

    Cover—page 117

    Student Directions—page 119

    Puzzle Pieces—pages 121 and 123

2. Reproduce the record sheet on page 125. Place copies in the left-hand pocket of the folder.

3. Place the envelope of puzzle pieces in the right-hand pocket of the folder.

Using the Center

1. The student takes the puzzle pieces from the envelope and puts them together to make a girl.

2. A correctly completed puzzle will show a girl.

3. Then the student completes the record sheet by drawing a picture of his or her own body and drawing a line from each body part name to that part on the drawing.

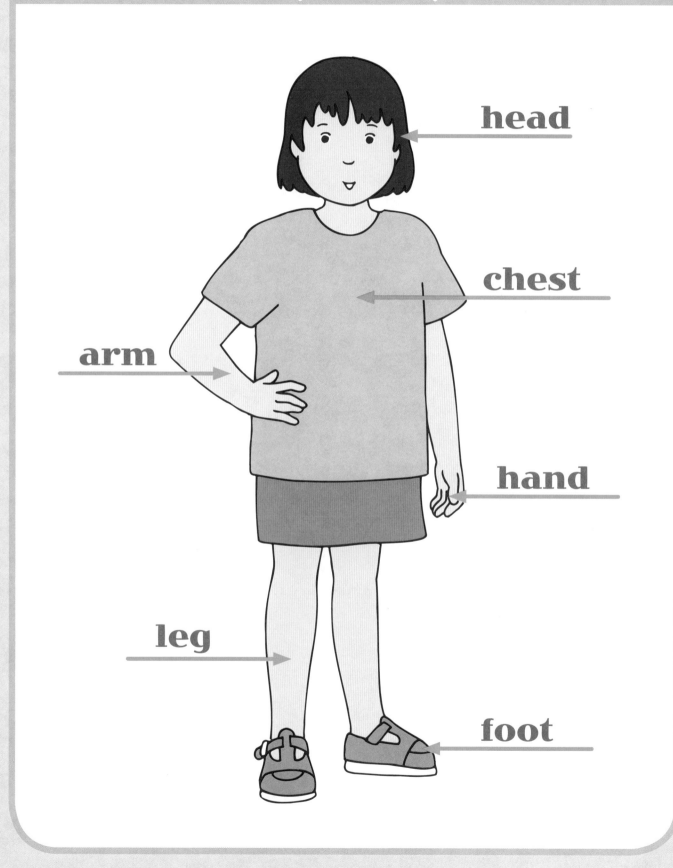

My Body

Follow these steps:

1. Take the puzzle pieces.

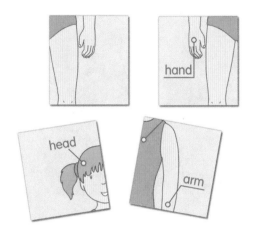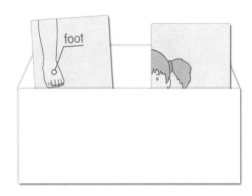

2. Put the puzzle pieces together to make a girl.

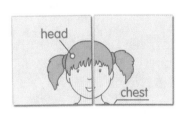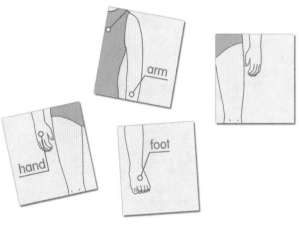

3. Take the record sheet. Draw a picture of yourself. Draw a line to make a match.

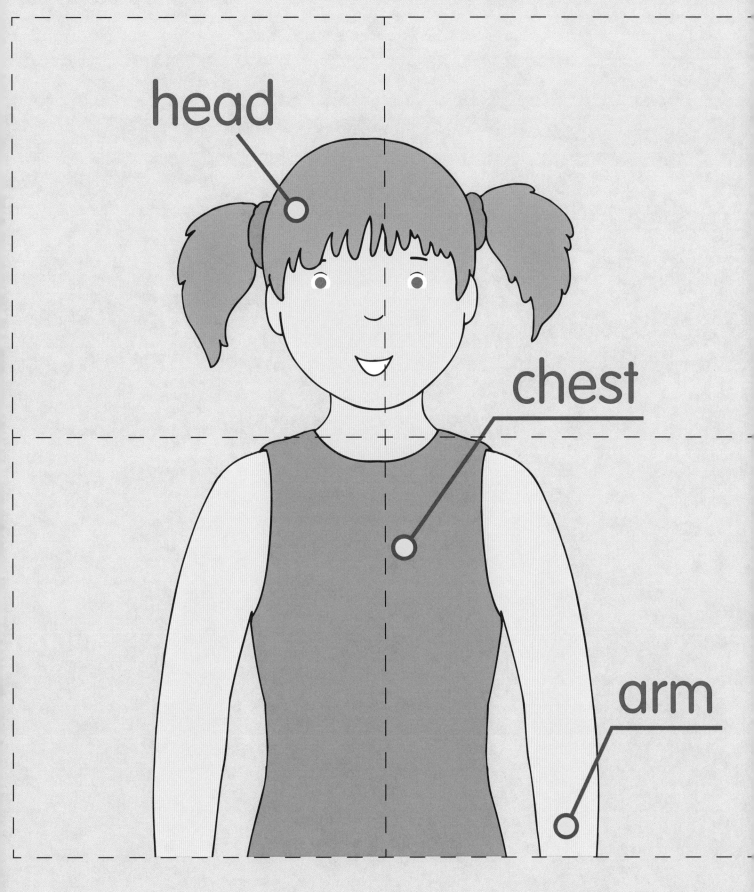

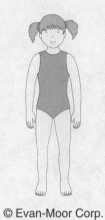
© Evan-Moor Corp.

© Evan-Moor Corp.

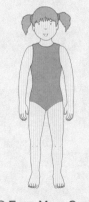
© Evan-Moor Corp.

© Evan-Moor Corp.

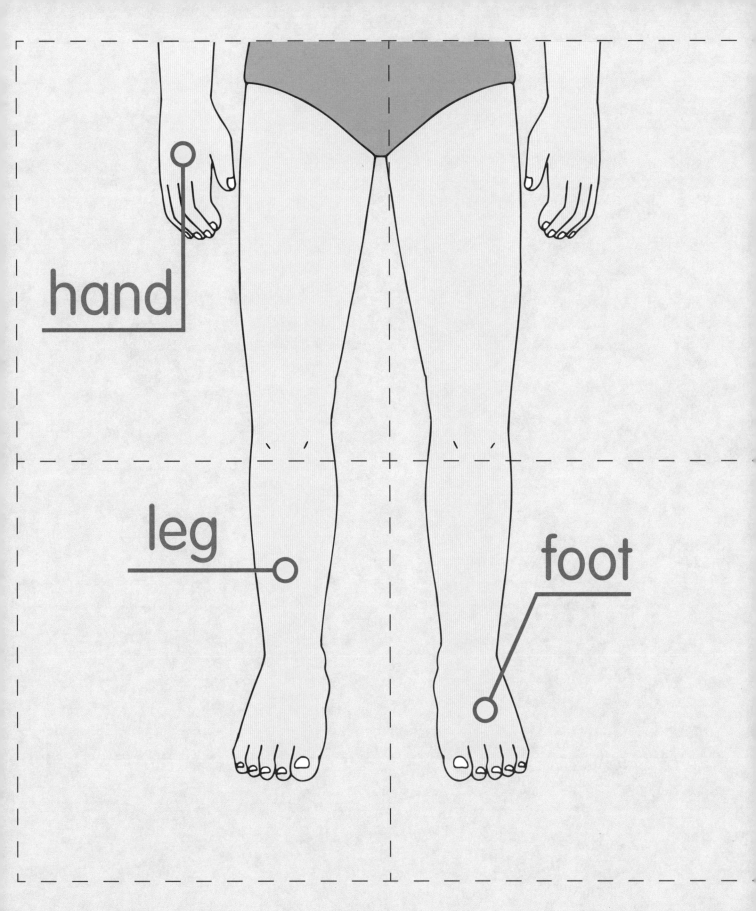

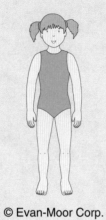

© Evan-Moor Corp.

© Evan-Moor Corp.

© Evan-Moor Corp.

© Evan-Moor Corp.

Name _____ Record Sheet

My Body

Draw a picture of yourself.   Draw a line to make a match.

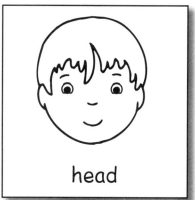
head

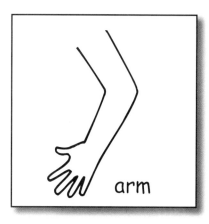
arm

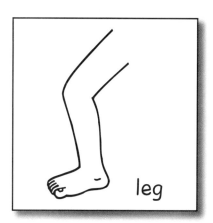
leg

What Will I Wear Today?

Preparing the Center

1. Prepare a folder following the directions on page 3.

    Cover—page 127

    Student Directions—page 129

    Sorting Mats—pages 131–135

    Clothing Pieces—pages 137–141

2. Reproduce the record sheet on page 143. Place copies in the left-hand pocket of the folder.

3. Place the sorting mats and the envelope of clothing pieces in the right-hand pocket of the folder.

Using the Center

1. The student takes the sorting mats and clothing pieces from the folder.

2. Next, the student dresses the three children on the sorting mats in clothing appropriate for the weather.

3. The student uses the self-checking clothing pieces to check his or her answers. The back of each piece has a symbol showing the type of weather in which it would be worn.

4. Then the student completes the record sheet by drawing a line to match each piece of clothing to a type of weather.

What Will I Wear Today?

What Will I Wear Today?

Follow these steps:

1. Take the mats and the clothing pieces.

2. Dress the children.

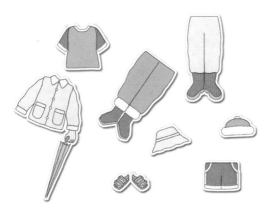

3. Take the record sheet. Draw a line to make a match.

A Snowy Day

A Rainy Day

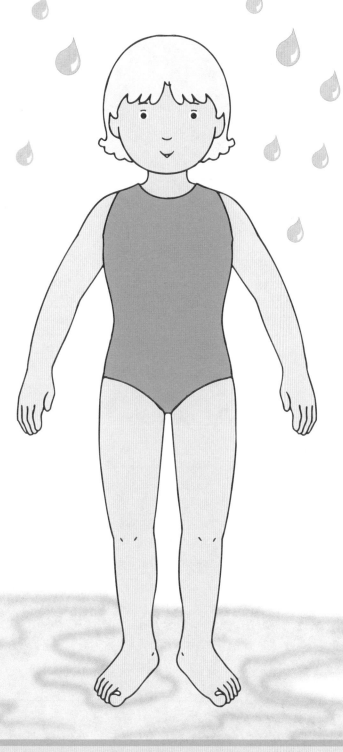

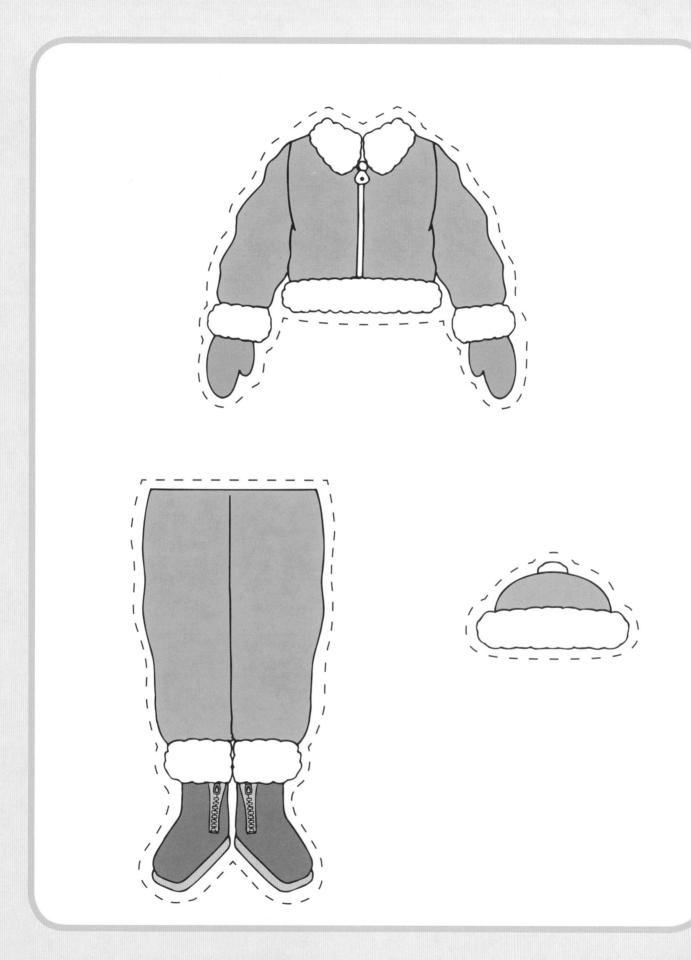

© Evan-Moor Corp.

© Evan-Moor Corp.

© Evan-Moor Corp.

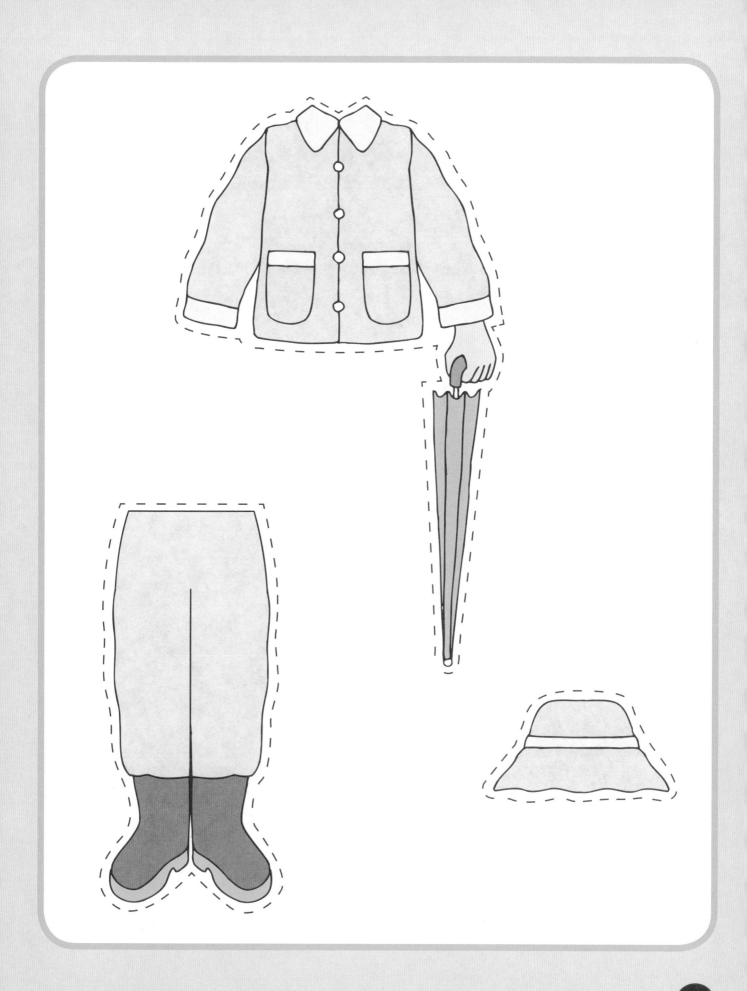

© Evan-Moor Corp.

© Evan-Moor Corp.

© Evan-Moor Corp.

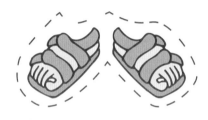
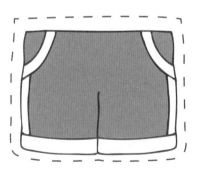

© Evan-Moor Corp.

© Evan-Moor Corp.

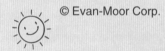

© Evan-Moor Corp.

Name _____   Record Sheet

What Will I Wear Today?

Draw a line to make a match.

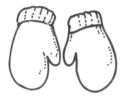

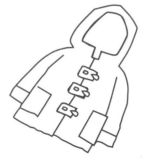

Animal Homes

Preparing the Center

1. Prepare a folder following the directions on page 3.

    Cover—page 145

    Student Directions—page 147

    Puzzle Pieces—pages 149 and 151

2. Reproduce the record sheet on page 153. Place copies in the left-hand pocket of the folder.

3. Place the envelope of puzzle pieces in the right-hand pocket of the folder.

Using the Center

1. The student puts the puzzle pieces together, matching each animal to its home.

2. The student uses the self-checking puzzle pieces to check his or her answers. Matching pieces have a picture of the appropriate animal on the back.

3. Then the student completes the record sheet by drawing a line to match each animal to its home.

Animal Homes

Animal Homes

Follow these steps:

1. Take the puzzle pieces.

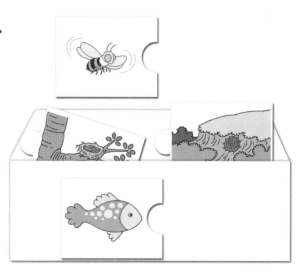

2. Match two pieces.

3. Take the record sheet. Draw a line to make a match.

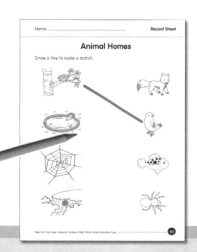

© Evan-Moor Corp.

© Evan-Moor Corp.

© Evan-Moor Corp.

© Evan-Moor Corp.

© Evan-Moor Corp.

© Evan-Moor Corp.

© Evan-Moor Corp.

© Evan-Moor Corp.

© Evan-Moor Corp.

© Evan-Moor Corp.

© Evan-Moor Corp.

© Evan-Moor Corp.

Take It to Your Seat—Science Centers • EMC 5004 • © Evan-Moor Corp.

Name _____   Record Sheet

Animal Homes

Draw a line to make a match.

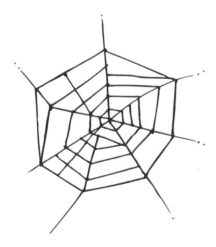   

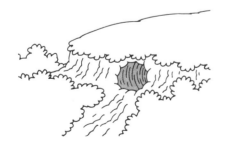   

Solid or Liquid?

Preparing the Center

1. Prepare a folder following the directions on page 3.

   Cover—page 155
   Student Directions—page 157
   Sorting Mats—pages 159 and 161
   Task Cards—pages 163 and 165

2. Reproduce the record sheet on page 167. Place copies in the left-hand pocket of the folder.

3. Place the sorting mats and the envelope of cards in the right-hand pocket of the folder.

Using the Center

1. The student sorts the cards into two sets on the mats—*solid* and *liquid*.

2. The student uses the self-checking cards to check his or her answers. All cards showing solids have an ice cube on the back. All cards showing liquids have a pitcher pouring water on the back.

3. Then the student cuts and glues the objects into the correct boxes on the record sheet.

Solid or Liquid?

Water is liquid.

Ice is solid.

Solid or Liquid?

Follow these steps:

1. Take the mats and the cards.

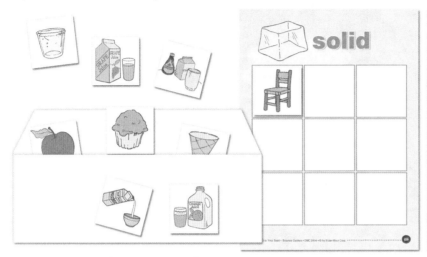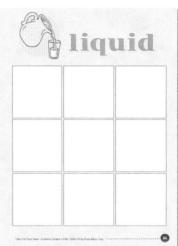

2. Sort the cards onto the mats.

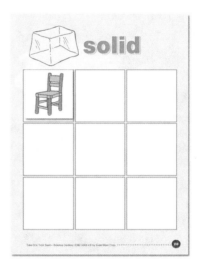

3. Take the record sheet. Color. Cut. Glue.

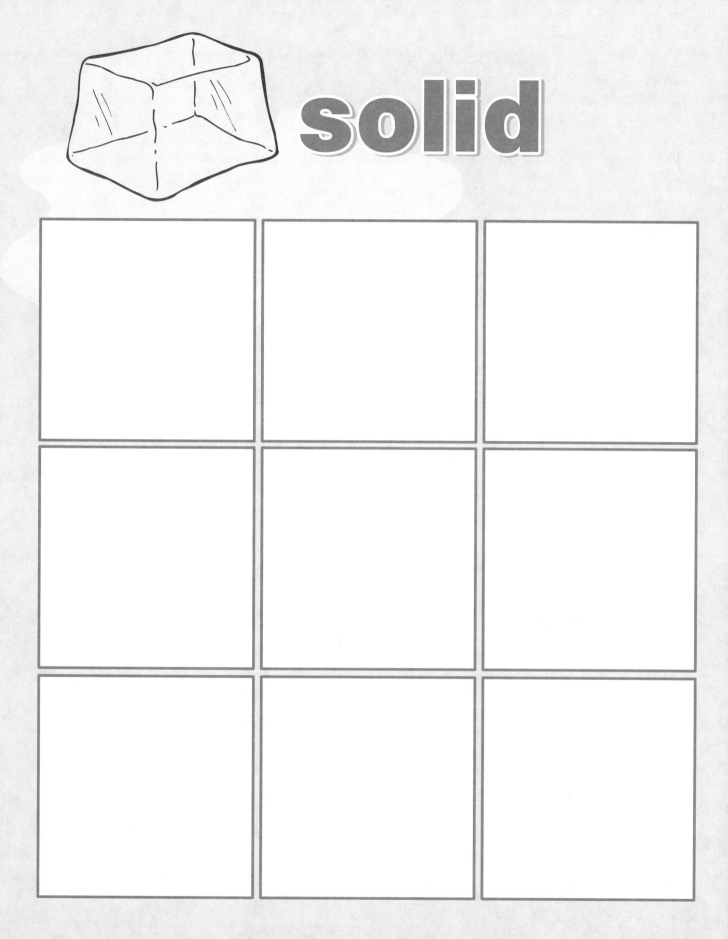

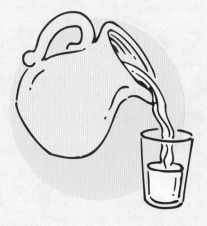 **liquid**

© Evan-Moor Corp.

© Evan-Moor Corp.

© Evan-Moor Corp.

© Evan-Moor Corp.

© Evan-Moor Corp.

© Evan-Moor Corp.

© Evan-Moor Corp.

© Evan-Moor Corp.

© Evan-Moor Corp.

© Evan-Moor Corp.   © Evan-Moor Corp.   © Evan-Moor Corp.

 © Evan-Moor Corp.   © Evan-Moor Corp.   © Evan-Moor Corp.

 © Evan-Moor Corp.   © Evan-Moor Corp.   © Evan-Moor Corp.

Name _____ Record Sheet

Solid or Liquid?

Color. Cut. Glue.

solid  **liquid**

glue	glue		glue	glue
glue	glue		glue	glue

Animal Puzzles

Preparing the Center

1. Prepare a folder following the directions on page 3.

    Cover—page 169

    Student Directions—page 171

    Puzzle Pieces—pages 173–177

2. Reproduce the record sheet on page 179. Place copies in the left-hand pocket of the folder.

3. Place the envelope of puzzle pieces in the right-hand pocket of the folder.

Using the Center

1. The student takes the puzzle pieces and puts them together to make animals.

2. The student uses the self-checking puzzle pieces to check his or her answers. The back of each set of pieces has a picture of the complete animal.

3. Then the student completes the record sheet by cutting and gluing the missing parts in place.

Animal Puzzles

Animal Puzzles

Follow these steps:

1. Take the puzzle pieces.

2. Put the pieces together to make an animal.

3. Take the record sheet. Color. Cut. Glue.

 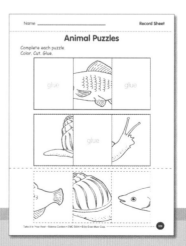

© Evan-Moor Corp.
 © Evan-Moor Corp.
 © Evan-Moor Corp.

 © Evan-Moor Corp.
 © Evan-Moor Corp.
 © Evan-Moor Corp.

© Evan-Moor Corp. © Evan-Moor Corp. © Evan-Moor Corp.

 © Evan-Moor Corp. © Evan-Moor Corp. © Evan-Moor Corp.

Name _____   **Record Sheet**

Animal Puzzles

Complete each puzzle.
Color. Cut. Glue.

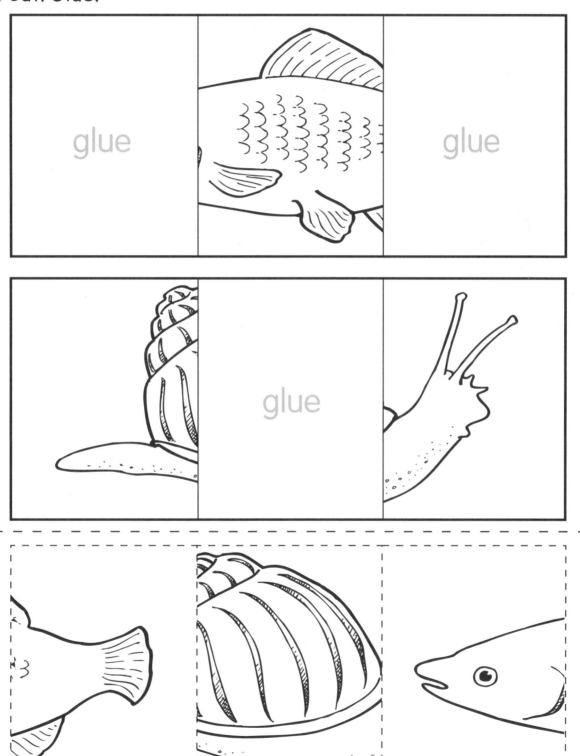

Sun, Earth, Moon

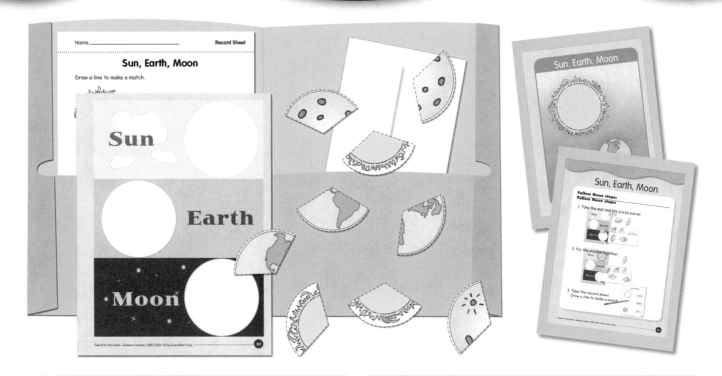

Preparing the Center

1. Prepare a folder following the directions on page 3.

    Cover—page 181

    Student Directions—page 183

    Sorting Mat—page 185

    Puzzle Pieces—page 187

2. Reproduce the record sheet on page 189. Place copies in the left-hand pocket of the folder.

3. Place the sorting mat and the envelope of puzzle pieces in the right-hand pocket of the folder.

Using the Center

1. The student takes the sorting mat and the puzzle pieces, and then puts the puzzles together.

2. The student uses the self-checking puzzle pieces to check his or her answers. The back of each set of pieces has the same color dot.

3. Then the student completes the record sheet by drawing lines to make a match.

Sun, Earth, Moon

Sun, Earth, Moon

Follow these steps:

1. Take the mat and the puzzle pieces.

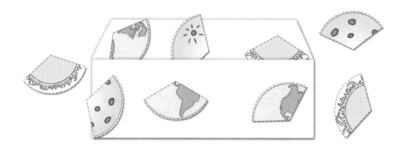
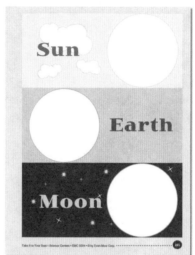

2. Put the puzzles together.

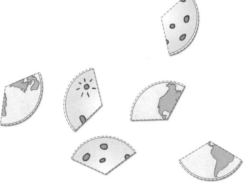
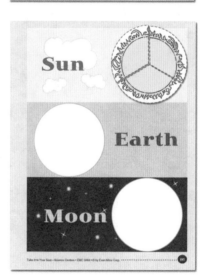

3. Take the record sheet. Draw a line to make a match.

Sun

Earth

Moon

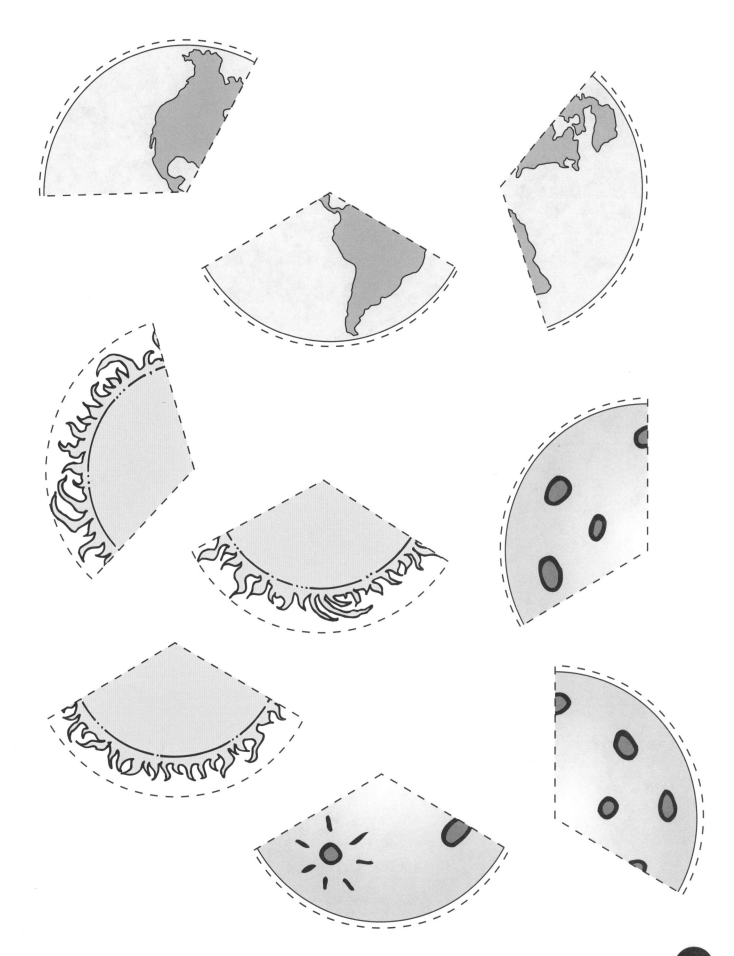

Take It to Your Seat—Science Centers • EMC 5004 • © Evan-Moor Corp.

Name _____   Record Sheet

Sun, Earth, Moon

Draw a line to make a match.

Answer Key

Page 13
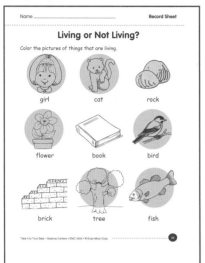

Page 51
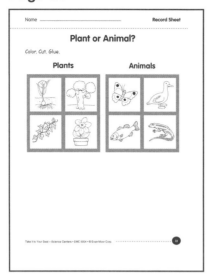

Page 81
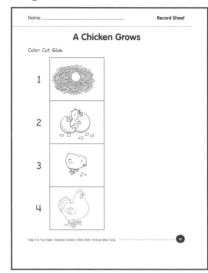

Page 25

Page 61
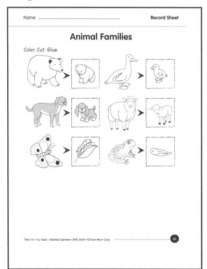

Page 95
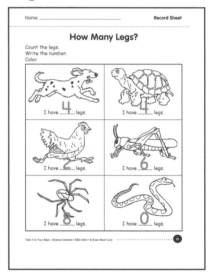

Page 37
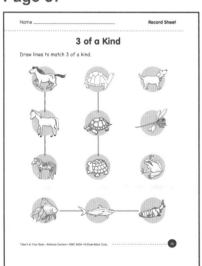

Page 71
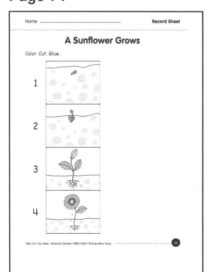

Page 105
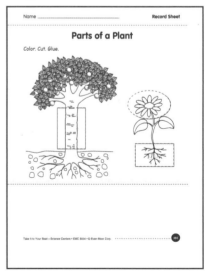

Page 115
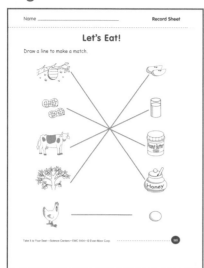

Page 153
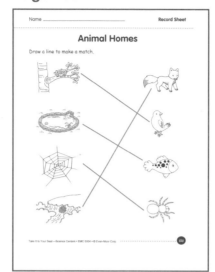

Page 189

Page 125
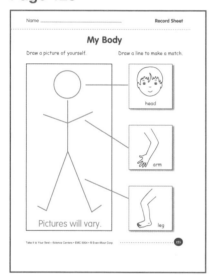

Page 167

Page 143
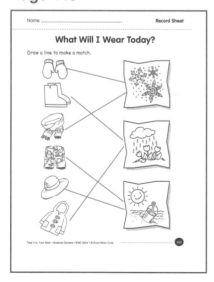

Page 179
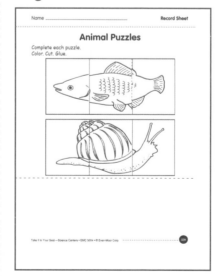

Take It to Your Seat—Science Centers • EMC 5004 • © Evan-Moor Corp.